Kenny,
Love ya man!

Exposure to Closure

The New Age of High-Tech Business and Marketing

By Stephen Sadler
CEO, President, Scate Technologies, Inc.

Published by ADAS Publishing, LLC, Clarkston, Michigan.

ISBN 978-0-9830161-0-6

SAN 859-9963

Printed in the United States of America

10 9 8 7 6 5 4 3 2 1

I dedicate this book to my wife Laurie, my children, and everyone else that has tolerated my intense passion and still stuck by my side. My eternal thanks.

Table of Contents

Foreword

It's Never Good Enough and It's Never Done

All businesses face challenges from time to time, but when the going gets really tough, such as the economic meltdown of 2009, it takes a special breed of entrepreneur to not only survive the storm, but to find opportunities to grow as well. Stephen Sadler (Steve) is that kind of entrepreneur, who sees obstacles merely as opportunities to discover what is on the other side.

You will never hear the phrase "good enough for now" cross Steve's lips. He only accepts solutions and ideas from others and himself that are complete, well done and above all, creative. Those that know him will say that he is unstoppable, and this attitude permeates every aspect of how he leads SCATE Technologies, Inc (SCATE).

Steve has an uncanny ability to combine the good ideas of his employees and his business partners with his own unique brand of thinking to produce fresh and original concepts for his company's products and services. Calculated risks are a necessary element of each and every day in Steve's world, and while he won't jump off a cliff without looking first, if there appears to be a way to make it work, Steve Sadler is not afraid to try.

Most true entrepreneurs will tell you that launching a new business is a high-risk, high reward adventure. Starting a self-funded business directly ties one's personal financial security to that of the business. As the income of the business ebbs and flows, so does one's personal income. Steve has faced down these issues multiple times as he has guided SCATE through the metamorphosis into a successful software development company.

Steve is well known in his community as a leader, innovator and as an involved parent. He can be found coaching girl's soccer teams several times a week on the local fields or playing ice hockey at the local arena. Speaking engagements at local schools and organizations are common for him. Steve is also very active in his church and volunteers at church functions. SCATE also sponsors community school events and sports teams, and has made a large software donation to many schools in its home state of Michigan.

Jeff Holth,

CTO, Scate Technologies, Inc.

The New Age of Business and Marketing

Social media marketing, pay-per-click, Twitter, Facebook, LinkedIn, search engine optimization… what does it all mean and why should you even care?

We all know that the economic meltdown of 2009 was tough for everyone. Even the seemingly invincible banking and insurance sector was almost brought to its knees. What was also interesting, and not so highly publicized, was how the recession also affected the marketing sector.

Many large marketing firms, especially here in the Detroit Metropolitan area, are now gone. It's not very often you see large marketing firms just close up shop, especially when marketing is so important to gaining exposure for new products and services.

Global competition is tough, so is it possible that companies can no longer afford expensive multi-million dollar marketing and ad campaigns? What about smaller companies, can they even afford to pop $10-25k on traditional marketing techniques every year?

So what if expensive, old-school marketing is partly responsible for the continued recession due to our inability to cost-effectively compete?

As with everything problematic, necessity is the mother of invention and on the forefront of our recovery is Social Media Marketing.

About My Book

Living in Michigan for the last 15 years, I know many people that have been laid off and are starting new businesses. I wanted to help by giving new entrepreneurs a few tips and tools to help them succeed in today's rapidly changing hi-tech world.

I was always fascinated by how Jesus taught through real-life stories called parables. I wrote this book in story format so you will experience first-hand how my wife and I started a business, and successfully and cost-effectively marketed a product, even in the most trying of times. The story starts back in 1983 and includes the 2003 inception and growth of Scate Technologies, Inc., the launch and continued success of Scate's popular social media technologies: Scate Ignite, IgniteCAST.com, Screentweet.com and Buztweet.com.

I hope you enjoy it.

Stephen Sadler

October 16, 2010

Exposure to Closure

The New Age of High-Tech Business and Marketing

Section I: Learning How to Lead and Succeed

The 80's - It Seems Like Only Yesterday!

Do you remember the 80's? Well I remember that era as if it was yesterday...big hair, two recessions, and a flock of seagulls.

I was 18, just graduated high school and was living on my own with not a penny to my name. It was August of 1983, and through a friend of my sister, I luckily landed a job working on the shop floor of a tool and die company in Chatham, Ontario, Canada. I was a posted welder making about $12 per hour, which was great money at the time. The work was physical but mindless, and for some reason I felt like I was back in high school. The guys were great but very immature and drugs were rampant.

It seemed every time I looked into the eyes of the co-workers who had been there over 20 years, I saw no future in their sight. Fear of being there until I retired made me literally hate every day I stepped foot into that building.

Consequently, in January of 1986 I took my first leap of faith and resigned from my job to open a Waterbed Store with a long-time friend in Goderich, Ontario, Canada.

"Finally I'm free and this is going to be the time of my life!" I said to myself as we drove up the beautiful coast of Lake Huron to open the new store.

Well, I was somewhat right and this turned out to be my first experience with "exposure to closure." Even before I became so-called "educated," I had a gift to see things in their simplest form. So to me, gaining exposure was all about making as much noise as possible, while closure was all about quickly making friends. This process seemed pretty simple to me and was a lot more fun than welding car parts together all day long.

I remember one Saturday sale in particular when we brought in over $25,000 of waterbed sales using radio spots and a station remote broadcast from our store. We even had a couple of friends stand at the road dressed up as the Cookie Monster and Big Bird to wave people inside. Due to the amount of customer traffic, we also hired two additional sales staff for the day to help close the hot waterbed deals. Selling seemed so easy and my partner and I actually turned it into competition.

Another simple marketing tactic was sponsoring an over-18 women's softball team. However, being single and twenty-two, the rationale for the sponsorship had little to do with marketing and more to do with meeting women. ☺

Now these marketing methods might not seem too radical for today, but back then, and especially in rural Canada, this was not very common. We were just a couple of dumb kids without fear trying new ways to have fun and socialize within the community...and it worked! But, like most new entrepreneurs, you need to lose it all before you become successful. Our lack of operational experience, coupled with not knowing anything about accounting and budgeting, caused the business to last only a couple of years.

☑ **Tip**: Entrepreneurship: Once this feeling of freedom bites you, you're stuck with it forever.

☑ **Tip**: If at first you fail in business, try again. Each attempt is a learning experience for you to build upon.

Getting Back On Track

In 1987, shortly after coming back from Goderich with my tail between my legs, I enrolled in university with the goal of getting a business degree. While still in my first semester, I met Laurie Benoit, my wife-to-be.

Laurie was a school behavior counselor at the time and was good at guiding people. The running joke was that she was the only one that could have controlled my behavior. In conversation one day she asked why I was taking business. I knew where she was going because I complained about it all of the time. The bottom line: I was taking business just because some of my friends were and she knew it. She also kept telling me that I had talents and that I should look for a job to start using them.

I felt she was just trying to get me to stop hanging out at her apartment, which in hindsight, was probably the truth. The crazy thing was that, to this point, I never really thought I had any real skills. I knew God had given me the talents for art, but how could these skills be the stepping-stone into a career? Besides most artists die alone and broke!

Shortly after the end of my first university semester, I started looking for a job where I could make money for school and gain more work experience.

Finding the Hi-Tech World in Rural Canada

I remember driving up to the Chatham unemployment office in March of 1988 and thinking to myself, "This is completely useless; there are never any jobs up here!" But to my surprise there was a job and it was almost exactly what I had been praying for, an entry level position in the Engineering Department at a large furnace manufacturer in Southern Ontario.

The new position required me to audit furnace components and enter part quantities into the bill of material (BOM) inside a material reporting program (MRP). The program ran on a large IBM mainframe computer and was my first real exposure to large computer systems. It really sparked my interest in the hi-tech computer world. The problem was, I knew very little about how they worked.

In 1989, I was fortunate to land a higher paying position as a CATIA Designer at an automotive parts manufacturer, also in Southern Ontario.

This new job entailed designing brake pedal brackets and car door mechanisms on CAD (CATIA). At that time, the CATIA software was running on a UNIX based IBM remote terminal. What I didn't realize was that this experience was allowing me to learn the inner workings of the UNIX operating system and learn how to transfer files through a token ring network. Little did I know how valuable these skills would become in the future!

Around the same time, I enrolled back in school and slowly start working towards my Bachelor's of Science degree at night.

☑ **Tip**: "Never say 'never'." I would not be where I am today if I had listened to that little voice telling me, "There are never jobs at the unemployment office!"

☑ **Tip**: It took me ten years to complete my degree. So if you're thinking about taking a night class, do it! Time will pass whether you do it or not and it's never too late to learn.

Learning from Bodybuilding - Is That Even Possible?

Since leaving high school, weight training had become a passion of mine. It was mindless and allowed me to pour stress into metal.

I wanted to compete in bodybuilding at least once, so in September of 1989 I entered a bodybuilding contest in Sarnia, Ontario. This was my first real experience being involved with any type of on-stage promotion. Even though I was one of the competitors, I was mesmerized by how many people came to see the event. I started to look at the types of marketing techniques the event organization was using to gain exposure. Funny, it turned out to be just like my days at the waterbed store...making noise in the community to attract attention.

I didn't realize until much later that bodybuilding was not just about lifting weights. It was about the WOW factor when you hit the stage.

Looking back, I now realize that it was just like marketing and sales. Each competitor had to have their own routine, a way of presenting their physique through choreographed movements to music. The routine was designed to bring the best out of your build. It also had to provide the audience and judges with the greatest amount of positive "exposure" while giving the judges reason to select you as the winner ("closure").

At the time, I was pretty upset that I didn't win. But, in retrospect, it was still worth every dry chicken breast and rice cake!

☑ **Tip**: If you let things happen and act on opportunities in a timely fashion, you will look back and see how your life fit so seamlessly together.

The 90's - Al Gore's Internet

What about the 90's? Al Gore had just invented the internet and everybody could pay their mortgage and even finance a second one.

In 1990, I was fortunate to land a Design Engineer position with another automotive supplier in Windsor, Ontario. I had a great relationship with my boss, the Director of Engineering, and he allowed me the flexibility to learn and grow. In 1992, he came to my desk and in his English accent said, "Steve you understand computers. Go downstairs and check out what's on the dock." So we went out to the plant and there stood a bright orange monster, seven or eight feet high.

"This is the future of welding, do you want to learn how to program them?" he asked. I quickly replied, "You bet!"

I was like a kid in a candy store and he placed me in charge of setting up the first robotic welding and assembly facility. After taking a programming course at ABB robotics, I quickly learned how to program large welding robots and control devices through input/output (I/O).

Again I was intrigued with technology, but the scary thing about these robotic monsters was that they could easily kill somebody if you didn't know what you were doing. This fear necessitated a high standard for programming code perfection that still influences my level of quality today; some of my developers now call me "pixel eye" because I want everything to be perfect.

When my boss left in 1995, I was promoted into his position as Director of Engineering. Shortly after, through a merger, the company became a much larger global corporation and my responsibilities increased significantly.

☑ **Tip**: Good leaders allow their employees to grow.

☑ **Tip**: If you're going to do something, do it right or don't do it at all.

How a Simple Fax Can Change Your Life

In 1995, I was sitting in my office when a fax came across my desk from an internet service provider located in northern Virginia. The fax read something like "Affordable Business Internet Service." Since I already had internet access at home through AOL, I was immediately interested.

"Maybe we could use the internet to cut data transfer costs," I thought to myself. Within a week, I signed up the company and we were off into cyber space.

It wasn't long before I realized that the internet had far more possibilities than just reducing data transfer costs. Having already been trained and certified in UNIX operating systems, I had a fairly good understanding of TCP/IP networking. However, I wanted to learn more about how the internet worked at the server and application level. After a significant amount of research, I realized that I needed to gain more knowledge about Apache web servers, DNS servers, HTML, CGI scripting (Perl), databases, and so on.

This new techno babble seemed like a daunting challenge, but with my robotic programming and CAD experience I felt it was time to jump on the "dot com" bandwagon and try something that couldn't kill you.

In 1999, I registered for several Linux courses at IBM in Toronto. For those of you that don't know what Linux is, it's the UNIX operating system knock-off that ran very nicely on x386 chip (or PC). This made Linux a very cost-effective solution for running web servers.

After completing these courses, I really began to understand where the internet was going. This intrigue seemed to refuel the entrepreneurial passion I had felt during my first business venture. However I had a dilemma; I was now making a good salary as Director of Engineering, and that would be tough to give up.

It's funny looking back because the company made the decision easy for me. They decided to create another level of management and I was totally opposed to it. My goal was then clear; I would start to design my own company. What I didn't realize was that my new company's launch date was coming faster than I could ever imagine.

A Rude Awakening

On March 6 of 2001, I was getting ready for work when I got a call from my sister. Her voice was shaky while she explained that our thirty-three year-old nephew (our other sister's son), had committed suicide. I lived with his parents when I was ten, so I was very close to him. Actually, my sister's kids were more like brothers and sisters. Needless to say, this tragedy really affected me. These are the kinds of life-changing events that open your eyes to how precious and short life is.

The tragedy, coupled with my discontent with the situation at work, resulted in a quick life-changing decision. After praying about it, I walked into the President's office and resigned from the company that had provided my family with security and wealth for over eleven years.

☑ **Tip**: Life is too short to stay in a job you hate or do just for the money.

Nothing Is Ever a Bed of Roses

I remember walking out the door of the company for the last time and thinking to myself, "Finally, I'm free again! But this time, I'm going to let God guide my direction, not money."

In April of 2001, I incorporated a company with a partner. Shortly afterward, I wrote the company's first business/marketing plan and thought to myself that it was going to be different this time. What I didn't realize was that my entire business plan was based on my successes from my previous employer. To make things worse, I graduated with a Bachelors of Science, summa cum laude, from Central Michigan University in August of 2001... so I now had a piece of paper that told me I was smart. Big mistake!

Sure, the business plan read well and would probably have sold a few angel investors, but I wasn't designing and manufacturing car parts anymore and I certainly wasn't selling waterbeds. I had moved into the technology world and was about to get an education in what it really means to run a business and compete in the hi-tech world.

The basis of my first marketing plan was selling training and support services to corporate customers in the field of Computer Aided Design, primarily CATIA. I knew the industry fairly well and it seemed pretty straightforward. What I didn't realize was that the world was about to change more than I could ever have been prepared.

All of us remember where we were on 9-11, and it affected millions of people in different ways. Personally, I experienced financial ramifications. Many of our clients drastically reduced their external services budget in a matter of months from that dreadful day. This left my company with a useless business/marketing plan and an empty training facility.

On April 1, 2003, my partner, who was the CFO at the time, drained our corporate bank account on payday and left.

This person was a very good friend so I could not understand what would bring someone to do what he did. I realized that entrepreneurship is not for the faint of heart. Even I, who had many years of experience directing a multi-million dollar engineering department, was not yet prepared for running even a small successful business.

You see, to lead a successful business requires humility, patience, strength, trust, faith, commitment, persistence, loyalty, and a whole lot more. Basically all the skills taught by either the

Bible or the military. So trust me when I say, "Business is a Battle."

Even though I personally never served in the military, I now understand and respect my father and both of my grandfathers for their service in both World Wars and the Biblical values they tried to instill in me when I was young.

☑ **Tip**: To build a successful company you need to practice humility, patience, strength, trust, faith, commitment, persistence, loyalty and a whole lot more.

☑ **Tip**: I could not run a successful business without my faith in God.

Have a Nimble Plan - Not a Contingent One!

I tried to understand what lesson I was being taught. After closing up the company in early 2003, my wife and I started a new company called Scate Technologies Inc. with an entirely new mindset. I now understood that our business issues were not the lack of a contingency plan, but were something much deeper. I realized that business and marketing plans, unlike engineering plans, cannot and should not be solid or even contain contingencies, but in reality they need to be fluid, scalable, and nimble enough to recognize and accommodate new opportunities in the ever-changing and uncontrollable world around us.

What you must also understand is that this was a real challenge for me, especially when I came from an engineering management background that required full control, measurement, and accuracy.

☑ **Tip**: Business and marketing plans need to be fluid, scalable, and nimble enough to recognize and accommodate new opportunities in the ever changing and uncontrollable world around us.

Running With an Idea

Now armed with my new flexible plans and mind-set, I started looking for methods that could help companies reduce costs. I had been teaching computer-aided design for years (specifically CATIA), so I was quite aware how costly it was to take CATIA classroom courses. I also realized that, since 9-11, training budgets had been drastically decreased and traveling for training was no longer an option for many companies.

I began researching methods that could drastically reduce the cost of training and certifying engineers. I hoped to identify value and solutions that Scate could provide to the professional engineering, certification, and training market.

With assistance from my employees, we attempted to break into the world of online training, or e-learning. My first question was, "How can we build these courses while still allowing them to be interactive?" Flash seemed like the logical choice, so we started down the path of building interactive courses... all from scratch.

After about six months of struggling with Flash, I discovered that what we needed was a real e-learning authoring tool and screen recorder. So we tested, Authorware, Lectora, RoboDemo and Camtasia, which all had strengths in different areas. The outcome was quite obvious; these applications were either too complex, not fast enough, or not robust enough for creating interactive online CAD software courses.

"If only I could combine some aspect of each application and then add a few of my own..." I thought to myself.

The problem was most CAD software changed so quickly that trying to keep the training content up-to-date was almost impossible. Even the video file size of the captured screen recordings with RoboDemo (Now Adobe Captivate) and Camtasia were just too large to stream over the internet. But more important was their lack in two key areas.

The first was that they did not contain closed captioning capability for the hearing impaired, which was critical for my clients. Second, any slight modification to the screen recording required a complete re-recording of the presentation from scratch.

Therefore, staying firm to my new nimble approach to business, I hired several computer science graduates from local colleges. Alongside many other projects, I tasked them to help me build a simple desktop software tool that would allow a user (me) to quickly capture the computer screens in sequence, add a pointing arrow, closed caption text, and record my voice.

About three months later, we had the tool! At the time, I called it RCDS for Rapid Content Development System. I was finally happy creating courses and broadcasting them to engineers and designers all over the world via the web.

☑ **Tip**: Look for methods that reduce costs and the corresponding market to sell your product or service.

It's Funny How Things Happen

After about a year of successfully selling the new CATIA courses online, I happened to be visiting a customer when they asked, "How do you create these interactive engineering courses?"

I quickly explained the process and showed them our little tool on my laptop. I was astonished when their reply was, "We want to buy it! How much does it cost?"

Dumfounded, the only thing that came out of my mouth was "What? Uh, we don't sell it."

They replied, "Why not!?"

☑ **Tip**: Be open, available, and nimble enough to jump on opportunities when they are presented. If you don't, someone else will!

You Better Be Prepared to Act

Making and selling commercial software was not in my business and marketing plan, so now what?

I had a company that wanted to buy my software that was not for sale. This was my first chance to execute my new nimble approach to business and grasp an opportunity.

Shortly after my conversation with the customer, the Scate development team and I poured our efforts into transforming RCDS into a commercially viable software tool. This is where exposure to closure begins.

☑ **Tip**: I don't know what your beliefs are, but if you're praying for opportunities, you better be available and ready to act on them!

Section II: Exposure

Get Exposed - It's Not a Dirty Word

Obviously, by now I was accustomed to switching gears. Making the move from directing an engineering department for a large automotive stamping supplier to training engineers worldwide through e-learning solutions was a significant transition and required a lot of effort. What I didn't realize was that my latest move into the software world added a whole new level of complexity and competition that was going to require an entirely new approach for sales and marketing.

You see, like most entrepreneurial newbies, I just figured that once our software was shrink-wrapped in a box and commercially released, we would have CompUSA or BestBuy knocking down our door wanting to sell our cool new software tool. Unfortunately, to my disappointment, that wasn't the case. Unless you were Microsoft, software was no longer sold in the store; people now downloaded it from the web, tried it for free, and then bought it online.

This placed all the marketing responsibility on us... a bunch of techies. How the heck were we going to compete against companies like Adobe and gain exposure for our product via the web?

First problem was, we didn't even have a cool name for the product and I was positive that RCDS would not be commercially successful. I decided to have a brainstorming session at the Scate HQ with some trusted friends. It went so well that in less than an hour, the product was named Ignite and the commercial software product, Scate Ignite, was born.

☑ **Tip**: Don't be afraid to get help from others. It is impossible to know everything and that's OK.

Launching a Product Is Not Cheap - Do I Need an Investor?

Having a business that ruined what I thought was a good friendship made me completely close-minded to taking on another partner for Scate. Also, the last thing I wanted was an outside investor influencing or controlling financial decisions at our new company.

In 2003, while teaching college programming courses at night, I met an individual that owned a restaurant called Joe Kools. His name was Dan Rea and like me, he grew up in Canada and still played hockey. These commonalities created an immediate connection. So you see folks, it's all about hockey!

Seriously though, after I told him about my little software technology company, he wanted to learn more about what we were building and how he could potentially help.

With my past experiences I told him, "I don't need any help!" but that was not the end of it. He actually traveled around with me for about eight months. I started to wonder, is this guy going to go home or what? Dan was probably one of the most persistent dudes I had ever met. But, I liked him and I was starting to trust his business advice. By the end of 2004, Dan became an employee and an investor. More importantly, he became a good friend.

I have met a lot of new entrepreneurs and inventors. If there is one problem they all seem to have it is how to gain exposure for new products and services. Usually the first thing that goes through an entrepreneurs mind is, "I need an investor to help fund a sales and marketing campaign."

The problem is, even if you had the investment money how would you spend it? Would you buy brochures, radio ads, TV spots, a cool website, pay-per-click ads on Google, or maybe hire a marketing firm?

☑ **Tip**: People are placed in your path for both good and bad purposes, but either way you will learn from the encounter.

☑ **Tip**: Don't be too hasty to take investment money.

Getting Completely Off Track

In March of 2005, Dan and I decided to take a nine-day trip to Coimbatore, India to present our Ignite software to five universities and a huge motorcycle parts manufacturer.

At the time we had just spent a significant amount of money building our new headquarters in Orion, MI. But with all the excitement about outsourcing, it seemed like India was the place we needed to expand.

Not only did this experience open my eyes to how other people live, it gave me a real appreciation for another culture. I remember driving down the roads in India within a sea of motorcycles and seeing families of 4 to 5 people on a bike clipping down the dirt road at 40 miles per hour.

After being treated like royalty and presenting to the universities, we decided to travel to a resort up in the mountains for the last three days of the trip. We drove for hours through wild elephant terrain and started our ascent up the mountains to Punar. We drove another four hours through the mountains before it started to get dark and rainy. The dirt road we were on was only one lane and had a 1000-ft drop to the left. Our driver thought it best to see if we could stay somewhere for the night. After pulling up to what seemed like a small, old lodge, we were escorted inside by an older fellow with one eye. (Until seeing the movie, Slum Dog Millionaire, I did not know why he and many others only had one eye.)

The lodge turned out to be an old English tea plantation and this fellow was the keeper. I took the back bedroom and Dan took the front one. As I stepped into the bathroom, I noticed a spider, about the size of my hand, directly under my foot. After taking photos of it, I chased it under the sink. Exhausted, I tried to go to sleep on an antiquated and musty smelling bed that unpleasantly reminded me of a hockey locker room.

When I woke in the morning, I had a bite mark on my right arm and a 104-degree Fahrenheit temperature. Our chauffeur decided to rush me down the mountain to see a doctor. After taking several types of medication, my temperature came down and Dan and I were able to fly back to the USA. As we were flying, I noticed that Dan had been bitten on his left cheek.

Upon arriving in the USA, my temperature started to spike again, this time hitting 106 degrees Fahrenheit. After a night of ice packs, Tylenol, and Motrin from Laurie, my temperature finally broke.

This time, leaving me with the worst fever blisters I've ever had!

Nine days later, I was driving to work when I got a call from Dan's wife. She was hysterical and yelling something about a car accident and a seizure. After she finally told me where Dan was, I drove directly to the hospital where he lay in a coma. The doctor explained that Dan had a brain tumor and they were going to operate. I told the doctor about our trip to India and the bite mark, but he didn't seem to be listening to me. When I went back to the hospital later that night, Dan's vitals were failing fast. After doing a quick search on my Blackberry, I found that they had him on two conflicting intravenous medications.

I rushed to tell the nurse and she asked me, "How did you know, are you a doctor?"

I replied, "No, I'm just a friend with internet access." She quickly removed the IV.

The next morning Dan was awake. I was so freaked out by the medication screw-up from the night before that I called Dan's brother Perry, and told him to get Dan out of that hospital. Perry agreed and quickly had Dan airlifted to the Mayo Clinic where they immediately diagnosed his condition as encephalitis and operated on the left side of Dan's head to remove an infectious mass.

Though they could not prove it, I believe wholeheartedly that it was caused by the insect bite on Dan's face.

Dan spent the next year regaining his memory and trying to pull his life back together. Even though I really liked the people, the food, and the culture in India, the time and resources might have been better spent in our own backyard.

☑ **Tip**: The grass is always greener on the other side.

Does My Marketing Really Work or Does My Product Suck?

Since the conception of Scate Ignite, we have tried almost every traditional marketing method known to man. As with many products sold on the web, Scate Ignite was not a big-ticket item. At around $200 per license, the volume of sales had to be high in order to turn a profit. In addition, our company was - and still is - a self-funded enterprise, so I cringed every time I authorized payment for marketing.

I remember spending $25,000 on brochures, another $15,000 on cool software boxes, $5,000 to $10,000 a month on pay-per-click ads, $100,000 on a fully automated e-commerce website, etc.

After about six months of spinning our wheels with standard marketing methods, I was getting a little discouraged (yes, I'm human!) and I actually started to wonder if maybe it was the product itself.

I thought to myself, "Maybe it needs more features or cooler GUI, or... blah, blah, blah." Obviously adding new features would require more investment and our current customers were certainly not asking for more. Then I remembered what my wife, Laurie, said to me earlier that week,

"Everyone that uses Scate Ignite raves about it, including Chrysler and Blue Cross Blue Shield. So why wouldn't everyone else love it, too?"

It just didn't make sense. I didn't know why we were struggling, and with my easy successes during the 90's, it was driving me crazy!

☑ **Tip**: Make sure you listen hard to your customers. They will tell you everything you need to know about your product or service.

☑ **Tip**: Don't jump to conclusions about your product or offering. Doubt is normal - get over it.

Pay-Per-Click/AdWords - Hmmm! Does It Really Work?

Before I start this section I want to preface it by saying that pay-per-click can work, but if you don't do it right you will pay BIG. Later, I will explain when and how to cost-effectively use pay-per-click advertising without taking an entire training course to figure it out!

When I look back over the last ten years, I remember creating many different Yahoo and Google ad campaigns. In addition, I consistently used Google's Keyword Generator Tool to tweak and find new related keywords and keyword phrases. After struggling through my first year of the pay-per-click marketing world, I came across a website by Perry Marshall and purchased his Guide to Google AdWords. I learned a lot about AdWords from this online book and I would highly recommend it. However, being of an extremely competitive nature, I was still unhappy with the performance of pay-per-click. Luckily, we were still doing well through our corporate sales channel, but I was adamant about cracking the code for creating more web sales. I knew there must be more to it because it was obvious that many other companies were selling products through the web. But, how could we?

There were, and still are, two main ways to come up first in a search on Google and Yahoo. The first method is to pay them through the per-click bidding process that places your ad on the side of a search page when a keyword or keyword phrase is typed into the search engine. I originally believed that the ad position on the page was determined by how much you bid against other bidders. But after trial and error, I noticed that this was not always the case. Later on you will understand why.

The second method is to optimize your website for search engines so you organically move to the top without paying for it. This obviously requires knowledge of how search engines work and an upfront investment that will pay dividends in the future if done correctly.

Most people start by paying for clicks because it is relatively easy to set up and they have minimal upfront cost. This is where companies make a big mistake and we were no different... at first.

Scate was spending an average of $5,000 per month on Yahoo and Google. Pay-per-click was driving clicks but definitely not at the volume required to produce profit for a product with a low price point. Where the process fell short was that these search engine providers did not (and still do not) provide you with enough information on how their system operates. So, a novice will lose a lot of money for pay-per-click exposure.

☑ **Tip**: Find a great Google AdWords training site at http://www.perrymarshall.com.

Search Engine Optimization

Refusing to believe that everyone was spending large amounts of money for pay-per-click ads, I began spending a lot of time researching Search Engine Optimization (SEO). I found that it required you to optimize your website around keywords or phrases. Simply put, it allowed your website's link (URL) to appear on search engines when a person looked for information, using specific keywords or keyword phrases.

It seemed too easy. I thought to myself, "We are a software development company; we should be able to figure this out." Again, I was in for a big surprise.

In addition to not implementing the SEO correctly on our websites, my rationale for doing SEO was wrong. You see, I was more concerned about not paying Google and Yahoo than I was about trying to develop a website that not only drove focused traffic, but also closed the sale.

Mr. Google - Who Is He Anyway?

Over the years I have heard many people say, "Google just doesn't like me, they're not letting me on the first page!" and I laugh. So before I continue on with my story, I just want to clear up two things:

A. There is no "Mr. Google" or "Mr. Yahoo" sitting in a data center somewhere in San Jose waiting to increase or decrease your ranking based on favoritism, unless maybe you're Youtube.com... but Google owns them anyway.

B. Search Engines are looking for four main things:

1. **HTML Structure** - Does your web page meet the technical requirements for proper indexing?
2. **Relevance** - Does your page content contain and relate to the keyword or keyword phrase?
3. **Traffic** - Are people coming to your site?
4. **Cross Linking** - Are other people linking to your site?

HTML Structure, It's Not That Complicated

After months of tweaking the main index page at www.scateignite.com and trying many "tricks" to get to Number One, we finally realized that everything we were doing was temporary and did not keep us on the first page. To do it right, the key was to give the search engine exactly what it wanted: clean, well-formatted HTML with no errors.

Tip: This is the best website to test your HTML, http://validator.w3.org/ before it goes live.

Relevance of the Content to the Keyword

In addition to HTML structure, the relevance of the content was and still is very important. Content on the web page has to contain the keywords or phrases you want to optimize. What a concept! Again, no tricks.

However, we had a problem; how do we optimize our website for every possible case in which a potential Scate Ignite customer might search? This is when we realized our current website, and especially the main index page, could not adequately explain all the possible use cases... 135 in total. More importantly, even if our main index page could explain every use for Scate Ignite, what keywords would Google and Yahoo index first?

Landing Pages Are a Definite Key

We decided that we needed to create 135 separate micro-sites, or landing pages, that would be designed and optimized to address all potential uses for Scate Ignite.

For example, Scate Ignite creates and posts social media to Twitter and Facebook. The last time I checked, we were ranked first for the keyword phrase "Create Videos for Facebook." If you look at the content located here...

https://www.scateignite.com/scate_ignite_home/how_to_create_videos_for_facebook.html

...you will notice that the title, the URL, the content and the tags all contain the words "Create Video For Facebook". Therefore, relevance is made and the search engine is happy.

☑ **Tip**: Make sure your landing page title, URL, content, and tags all contain the keyword you want to optimize.

XML SiteMap - Say What?!

Now we had all these pages on our website that were fully optimized to answer questions, but nothing was happening. What were we missing? We had done everything right, but Google and Yahoo still did not know the pages even existed.

You see, most search engines use robots that sit on your website and scan for new content. Each webpage is then indexed based on the criteria we discussed earlier. We soon discovered that if the robots are not pointing to the new landing pages, it will take a long time to get indexed into Google and Yahoo.

To become indexed by the search engines, we created XML sitemaps. I'm sure many of you have heard the term "sitemap" from a user's perspective. Basically, it is a listing of web pages contained within your website that provides quick navigation to the page.

An XML sitemap, on the other hand, allows the search engine robots to easily find all the web pages you want indexed. This is simply a file you call sitemap.xml that is placed in your website structured in Extensible Markup Language XML which can be found here:

http://www.sitemaps.org/protocol.php

The file must contain all the URL links to every webpage and landing page on your site.

WebMaster Tools - Proper Submission!

Now we had our new sitemap.xml on ScateIgnite.com, but how did Google or Yahoo's robots even know it existed? The answer to this was actually pretty simple. We just submitted it to them.

For Google, we needed to create a Webmaster Tools account. Once inside our account, I added our site. I was then prompted to verify that the site was mine by adding a meta tag in the heading section of our index page at scateignite.com. I know it may sound complicated, but it was a simple, copy-and-paste procedure. Once the verification code was in place, I simply returned to the Webmaster Tool account and clicked the "Verify this site" link next to the site name, ScateIgnite.

Once verified, I entered the site page within the Webmaster Tools account and clicked "site configuration" > "sitemaps." I now needed to submit my XML sitemap by pasting the URL on the sitemap https://www.scateignite.com/sitemap.xml field and I was done. Sorry to get too complicated, but this step was critical for us to get www.scateignite.com submitted into the Search Engines.

Yahoo and Bing worked very much the same. (You can use Yahoo's Site Explorer located at:

http://siteexplorer.search.yahoo.com/.) The main difference at Yahoo was mainly terminology where sitemaps are called "feeds" and verification is called "authentication." Otherwise, the process was almost the same and at the time of this publication, it still is.

To submit to the hundreds of other search engines worldwide, we used WebCEO at http://www.webceo.com. It is very easy to use and works great.

☑ **Tip**: The more frequently you add or change links in the XML sitemap, the more frequently the robots will index your content.

Traffic and Cross Link - Sounds Like The Ticket

Our landing pages worked very well. But, to stay at the top of the search engines, we obviously needed traffic. Unfortunately, this is a catch-22. In my experience, the traffic comes after the pages are indexed properly.

Cross linking, when other websites have back-links to our landing pages, was also very important to our SEO. The more back-links we had, the better our ranking. To help facilitate cross linking of the landing pages, we included a social media bookmarking/sharing widget on the site. This allowed other people to easily post, bookmark, or share our landing pages on Digg, Delicious, Twitter, Facebook,and hundreds more.

☑ **Tip**: AddThis and Gigya are excellent social media bookmarking and sharing widgets that you can easily embed on any webpage. You can find them at http://www.addthis.com/ and http://www.gigya.com/

Google Pay-Per-Click and Landing Pages - Complementary!

Now you are probably wondering, what does a search-engine-optimized landing page have to do with pay-per-click? Wasn't my goal to stop paying Google and Yahoo?

Well, yes and no. If you're going to spend money, you want to spend it wisely.

When we started our SEO landing page campaign for Scate Ignite, we did not want to completely shut off Google and Yahoo until the natural search engine traffic increased. So, I created 135 keyword-specific pay-per-click ads that linked directly to each of the landing pages we just created.

To my surprise, we found that the price we were paying per click dropped drastically. We learned that the reason was, once again, relevance. The keywords and content in the ads now matched exactly to content and keywords on the landing page. Amazingly we were now paying $0.10 per click instead of $5.00. We also found that once we started paying Google and Yahoo to link to their ads, our landing pages seemed to get indexed for SEO almost immediately. Again, amazing!

Tip: Make sure the keywords and subject of Google and Yahoo pay per click Ads contain the keyword for which the landing pages have been optimized. This will lower your bid cost!

Social Media Exposure - It's Definitely Not a Fad

If you were to ask me what social media was five years ago, I would have looked at you like you had three heads. Since 2006 I have been fascinated with the rise of web applications such as LinkedIn, Facebook, Twitter, Digg, Delicious and YouTube.

At the official Twitter developer conference, "Chirp", Twitter revealed the following stats as of April 2010:

- Twitter now has 105,779,710 registered users.
- New users are signing up at the rate of 300,000 per day.
- 180 million unique visitors come to the site every month.
- 75% of Twitter traffic comes from outside Twitter.com (i.e. via third party applications).
- Twitter users tweet an average of 55 million tweets a day.
- Twitter's search engine receives around 600 million search queries per day.

* Of Twitter's active users, 37 percent use their phone to tweet.

The Facebook pressroom at http://www.facebook.com/press/info.php?statistics contains their stats. Here are the stats as of September 2010:

* More than 500 million active users.
* 50% of active users log on to Facebook in any given day.
* Average user has 130 friends.
* People spend over 700 billion minutes per month on Facebook.
* More than 30 billion pieces of content (web links, news stories, blog posts, notes, photo albums, etc.) are shared each month.
* There are more than 150 million active users currently accessing Facebook through their mobile devices.

Conclusion: It doesn't matter how you slice it folks - THIS IS WHERE CONSUMERS AND BUSINESSES ARE HANGING OUT.

The Rise of ScateIgnite and Ignitecast.com

Ironically, my own software tool, Scate Ignite, and its broadcast mechanism, Ignitecast.com, had been creating social media since 2004... I just didn't know what to call it.

When we hired our current CTO, Jeff Holth, in 2005, he asked me what to call the published output from Scate Ignite. "Should we call them presentations, videos, demos, FAQs, e-learning tutorials, video podcast, screencasts... what!?" he asked. Meanwhile, one of my sales reps had been calling them Ignitables. The name stuck.

We didn't realize that we were positioning ourselves to ride the largest wave since the rise of the internet in the late 90's.

In 2007, we were trying to provide our customers with a quick and easy way to broadcast and track the published files (Ignitables) made with Scate Ignite. Together with three of my developers, we decided to build Ignitecast.com to help our customers share their media.

We finally knew what we had been creating for years. It was not just e-learning courses and presentations. It was social media... and, it was easy to make and easy to share with Scate Ignite and Ignitecast.com.

☑ **Tip**: Create marketing presentations (Ignitables) with Scate Ignite and broadcast them on Ignitecast.com and YouTube.

Using Social Media to Gain Exposure

One of the key things I learned while building Ignitecast.com was that most social media sites are search-engine-optimized. If they weren't, your content would never go anywhere, especially not in real time. For this reason, when you put video on YouTube or an Ignitable on Ignitecast.com, it is almost instantly indexed into Google, Yahoo, and Bing.

Therefore, everything I explained in the SEO section of this book directly applies to all the social media technology sites you use on a daily basis.

Consequently, we started using Scate Ignite to create social media presentations about Scate Ignite's own capabilities and then broadcast the Ignitable out on Ignitecast and YouTube. It was kind of "the mirror in the mirror" thing; I actually thought a black hole was going to appear and devour us at one point. However, this worked so well that Scate Ignite and Ignitecast started to show up all over the web and the snowball was off and running.

☑ **Tip**: Post Ignitables on Twitter, ScreenTweet, and Facebook. Bookmark them in Digg and Delicious.

Full Circle -
Those Darn Landing Pages are Back Again!

You know, it's funny how everything seems to come full circle if you wait long enough. After Ignitecast.com had been growing for about a year, we decided to overhaul the landing pages on the ScateIgnite.com site. I wanted to add a little more pizzazz, so instead of just using static pages, we started to embed the Ignitables on the landing pages from Ignitecast.com.

The results were incredible; not only did this produce far more Scate Ignite downloads, it was an excellent showcase of what Scate's Ignite technology could do... rapidly create and share social media with many in minutes.

☑ **Tip**: If I haven't emphasized this enough already, here I go again... Landing pages are *critical* to the success of your website!

Is Twitter Really the Game Changer for Business?

I still remember the day Jeff Holth, our CTO, asked me to create a Twitter account. Like many of you, I already had a Facebook account, a LinkedIn account, a Digg account, blah blah blah! The last thing I wanted was another site to maintain, but reluctantly I created one anyway.

Within about a month, I had about five hundred followers telling me about things I could have cared less about. At that point, I was fully convinced that this technology was interesting but completely useless.

One night in early 2009, Jeff Holth called me on my cell phone and blurted, "You need to buy the ScreenTweet.com domain, it's going to be huge!"

I hung up on him. ☺ But, after thinking for a moment, I called back to see what he was talking about. "Checkout Twitpic.com," he said. "They are getting tons of traffic."

"So?" I asked, not being one to build things that have absolutely no mechanism to generate revenue.

In an attempt to remain nimble, we scaled down a version of our ignitecast.com technology the very next day and by the end of the week, we launched Scate's new marketing engine... ScreenTweet.com.

ScreenTweet.com looked very promising and the amount of traffic it was getting was far larger than we anticipated. More importantly, ScreenTweet.com was the least expensive venture we had ever made and was generating ad revenue right out of the gate. Incredible!

☑ **Tip**: Sometimes it's the simple things that make you a success!

Many Tweets and Many Twitter Accounts

Shortly after experiencing the power of Twitter for creating a fan following, I started to realize that Twitter was an easy-to-use multi-directional broadcasting system for social media.

From a business perspective, I was starting to see far more potential for Twitter than any of the other leading social sites.

For example, Facebook was, and still is, geared around friends and family; unless you sell cars or MLM, you are definitely not selling to this crowd. LinkedIn was (and is) so locked down that it had become an online Rolodex of people I met. But Twitter was all about open communication with people, organizations, and businesses that I'd never met. These were people that were looking to network, buy, sell, discuss, comment, and even complain.

Tweets Are Like TV and Radio Spots - Say What?!

This is the part that gave me a headache, at first. But trust me and read carefully because this is very important to understand.

I noticed that if I posted a Tweet (a.k.a. micro blog) it would quickly be buried by other Tweets in the feed in a matter of seconds. Actually the Tweets seemed to have a very short shelf life of about four to eight seconds on the Twitter public feed.

Trying to think in simple analogies, this brought me to the conclusion that Tweets are like TV or radio spots. For example, no one buys one TV or radio spot, because the chance of someone tuning in during its broadcast would be very slim. However, if you buy a block of TV or radio spots that are played periodically, your chances of gaining exposure drastically increase.

The same applied to Tweets. Twitter was (and is) all about Tweeting and Tweeting often. Tweeters (Twitter Users) wanted to hear from you all the time. This was a change in mindset that troubled me, at first. "Isn't this spam?" I thought to myself?

Then I realized that spam is when you trick people into doing something and don't give them a way out. Tweeting a message or business tip is not spam if you're honest about what you're saying and the embedded links go to a relevant site. This is totally different from unsolicited email you get every day. Tweeters are free to follow and un-follow you at anytime, so it is definitely not spam.

I began generating and posting many different Tweets several times per hour with different messages containing shortened links. These links then directed Tweeters to the same landing pages I created during the SEO phase.

Wow! What an excellent way of communicating a message to millions of potential clients, qualifying them, and then funneling them to a page that contains a call to action. This worked so well that we reduced and, in some cases, completely stopped spending money on other forms of traditional marketing solutions including pay-per-click (i.e. Google AdWords).

Plato was a genius when he said, "Necessity, who is the mother of invention." Like many of the software products created at Scate, this internal social media marketing approach turned into another commercial technology/service which is now called BuzTweet.com.

☑ **Tip**: Tweets are like TV or Radio spots and require quantity, variety, and frequent broadcasting.

☑ **Tip**: If you do social media right, you can drastically reduce and in some cases completely stop spending money on other traditional marketing approaches.

☑ **Tip**: If you combine social media with Buztweet.com, you have millions of potential clients just sitting at your fingertips!

Why Can't I Understand This New Technology?

One of my sales reps that came from the "old school" of sales was having a tough time adjusting to all these new social media sites and online networking technologies such as Twitter.

This fellow was one of the best relationship-builders I had ever met, but just could not grasp how the electrons on the screen could turn into dollars. So I tried a different approach to help him understand this new powerful technology in his own frame of reference.

First I told him to think of Twitter Tweets as "word-of-mouth" and Twitter Followers as "potential clients."

"This seems simple," he said, "but how do I sell to them, many of whom I have never met?"

My answer was simple, "Search on Twitter for people with a common interest in what you're selling and start following them, read carefully what they're saying and then start conversing with them."

This is like eavesdropping at a conference to find someone with common interests. All sales people do this. The key here, and where Twitter differs, is that you don't have to go anywhere to do it! Millions of people are sitting at your fingertips and "TELLING YOU WHAT THEY WANT IN PUBLIC!!!"

☑ **Tip**: If new technology scares you, try to compare it to something you know.

Some People Can't See the Forest Through the Trees

I recently had a conversation with the owner of a marketing company whom I respect, but who I would consider "old school" in the realm of marketing methodology. He argued that social media is not another broadcast marketing channel. Shortly after the discussion, I read an article by the "Senior User Experience Researcher" from a prominent search engine company, and it seemed to reinforce what he had said.

At first I was annoyed, because I know that millions of companies are using Twitter and other social media sites for marketing and customer service. Then it dawned on me why these traditional marketing firms were against social media becoming a new channel.

They were afraid of it, and rightly so!

Social media marketing is a threat to how they currently generate revenue. The great thing about the free market system is that buyers will determine how a new invention or technology is utilized and evolves. Opinions do not generate profitability.

RESULTS DO!

So you see, in reality it doesn't matter what anyone thinks. People will use a technology when it helps them to be successful and is providing ROI (Return on Investment).

I am sure many of you would agree that many current marketing methods fail to provide ROI and this is why you are still looking for a better, cheaper, and more transparent alternative for marketing.

One thing is for sure, social media is here to stay! It definitely seems to be the next evolution in marketing technology just like the newspaper, TV, radio, and internet.

☑ **Tip**: Opinions do not generate profitability and revenue for business...RESULTS DO!

Section III: Closure

Closure. OK. What's That All About?

As you can see, exposure can really help your business, but exposure alone does not make a business successful.

As a serial entrepreneur, there are two things that come to my mind when someone tells me they are starting a new business.

1. What is the value proposition? In basic terms, why would somebody want your product?
2. How are you going to close profitable deals? In basic terms how are you going to make money for your family?

Be Aware of Changing Value Propositions

Most businesses think they know what their value proposition is; however, putting it into words can sometimes be far more difficult. The easiest way I accomplished this was to let the customer define our value proposition.

For example a value proposition of a product or service can change without the business even realizing it. When we first launched Scate Ignite, our value proposition was the cost-reduction of training through the rapid creation of e-learning. However, through time and the emergence of social media, Scate Ignite's value proposition changed.

This might not seem like a big deal because e-learning can easily fall into the bucket with social media; however, if we hadn't changed our message, we would have missed out on a far larger market than before. So, after asking many customers what they used our product for and what value it provided, it was relatively easy to re-position Scate Ignite for the new rising market.

Now, Scate Ignite's value proposition is the cost-reduction of marketing through the rapid creation and sharing of social media. Let's compare both value propositions...

The first:

The cost reduction of training through the rapid creation of e-learning.

And, the second:

The cost reduction of marketing and training through the rapid creation and sharing of social media.

It's amazing how changing a couple of words created an entirely new message that allowed us to attract and close a much larger market of users. Since the change, Ignitecast.com now has branded sites where large companies are embracing social media for their own corporate communication, global training, and social media initiatives. This might never have happened if we had not taken the time to listen to our customer and change our value proposition.

☑ **Tip**: Listen to your customer. They will define your value proposition.

Importance of Closing Only Profitable Deals

I remember a good friend, whom I highly respect, say something like this:

"If you can make profit on the job then your price is good."

He didn't say what percentage of profit; he just said "profit" and that has stuck with me for years.

I've been asked many times by companies to build applications for a fixed price and timing that they defined. I graciously turn that work down. Since starting Scate, I have not taken on any job that we could not make a profit on. As a business owner, you should not either. Usually, when you take on a losing job, the project and the business relationship go down the tubes really fast. Go figure.

☑ **Tip**: If you can make profit on the job, then your price is good.

Defining a Price

We all know it's a recession and frankly I am sick of hearing about it in the media. Bad news creates fear in people and the first thing businesses do is start second-guessing their pricing structure.

In many cases you see companies taking losses just to get the work, or dropping the price of their product to compete. This is pretty simple math. If you take on or sell a product or service for a loss, you will have to do one of following:

1. Lower your overhead.
2. Provide less product or service.
3. Eat the loss.

Many companies tend to go with eating the loss and this is usually the start of bankruptcy.

The goal is to define a pricing structure so that you are happy when you sell. Your price needs to include the cost of goods or services, cost of sales and marketing, plus profit. Never feel bad about making a profit: that is what business is all about.

Now if you can't compete, then that's another story. To compete, you need a better mousetrap, not a cheaper price. Price can only change if you find more cost-efficient methods that lower your cost of goods and services or the cost of sales and marketing. Profit should remain the same unless you want to aggressively win a bid. Either way you're operating in the black and that's where you want to be.

☑ **Tip**: Price should only change if you find more efficient ways to lower your cost of goods and service or the cost of sales and marketing.

The Call to Action and the ASK

Let's go back to the landing pages. If you remember, every landing page contains a call to action that should be very prominent on the page and should have a direct ASK. For example:

"Click Here to Buy"

or

"Click Here to Contact Us"

It sounds relatively simple, but I see many landing pages that provide many other things to do, when in reality, if you want the page to close a sale, then you had better not let the visitor do anything but buy, contact you, or learn more. Anything else is superfluous and will dramatically increase your bounce rate and decrease the performance of the page.

☑ **Tip**: Provide visitors of your landing pages with only one or two call-to-action buttons. Remove every other button or link.

Importance of Follow-Up

One of the areas I have seen businesses fail in is the area of sales follow-up. For example, if your landing page call to action is "Contact Us" or "Download Now" then you had better follow-up by phone, email, or social media immediately or you have lost the sale. Most people on the internet buy on impulse, depending on the type of product, of course. So if you don't contact them and provide them with the value proposition while they are in the buying mode, then forget it. It's too late.

But with the rise of social media, you can easily follow-up and communicate with hundreds or thousands of people that are interested in your business. For that reason, the key area we focused on when we built Buztweet.com was the ability to rapidly monitor and communicate with people that "retweet," "mention," or "direct message."

☑ **Tip**: Most people on the internet buy on impulse, so follow-up immediately, don't procrastinate.

Constant Contact

There is actually a company called Constant Contact and we actually use them, so here is a plug for them...

http://www.constantcontact.com/index.jsp

Just like the name, their service is good 'ol personal contact and timely follow-up, combined with consistent touches via email.

Ever heard the saying, "Out of sight out of mind?"

Well, let me tell you, it's true. If you're not consistently in your customers face they will forget you faster than the shelf life of a Tweet.

I am not talking about spamming people here. What I'm talking about is providing all your contacts with news, updates, or special promotions that they might be interested in and want. If you give them a way to opt out, everything will be cool.

This is one of the reasons that I like using Constant Contact; they do not allow spamming and have a great reputation in the email marketing industry.

☑ **Tip**: Constant Contact is a great way to consistently follow-up with your contacts and customer base.

The Power of Free

It's strange how many companies have become successful by giving things away. For example, Google and Yahoo give away tons of services for free.

Everyone loves free stuff, but as a business owner, you really have to be careful. Several years ago I heard one of our customers say, "I don't need to buy that software, I'll just get it for free over at Company X."

Well, he was right and Company X was giving it away for free. It wasn't exactly the same value, but it was definitely free. It turned out that Company X did not care two cents about their free software service. To them it was just a mechanism to create a community and then market to them for millions of other products.

Sure, I wasn't happy because I thought it devalued our product. To my surprise, however, it actually worked in reverse. Not only did it help increase our sales, it also helped to define a market that was non-existent, thus increasing our customer base.

☑ **Tip**: Giving things away for free can definitely help with sales, just be careful not to de-value your product or service.

Swimming Next to Big Fish

In the middle of 2009 we took the same approach and started to create a free service to help create our own community. In August of 2009 we launched ScreenTweet.com and it was the first free service Scate offered. Because it was tied to Twitter, it immediately generated thousands of users in just weeks and has steadily grown ever since.

Now I have been asked, "Why did you create ScreenTweet and how are you making any money from it?" The answer was simple. If you drive large amounts of traffic you will make money through advertising on the site.

Google AdSense is one way to embed advertisements on your website and then get paid by the number of clicks on those ads. However, there are many others companies that pay by the number of "impressions" (the number of times your ad is visible to a user). Obviously, advertisement is not just about clicking; it is also about brand recognition and many companies will pay you big if your site is getting enough traffic.

For us, ScreenTweet.com became a nice little resource of ad-generated revenue. But, more importantly, it gave Scate a lot of word-of-mouth exposure in our local community (Orion Township,

MI and the surrounding areas). People love hi-tech, and Silicon Valley does not have a patent on that!

☑ **Tip**: You don't have to be in Silicon Valley to be creative, folks. Just apply what you know to your dreams.

The Importance of Measurement - Google Analytics

One of the most important aspects in performance is measurement; measuring what landing pages are driving the most traffic and what landing pages have the highest closure rate.

Google Analytics provides some very valuable tools for tracking statistics and goal completion of your website. It seems complicated at first, but after using it for a while, it will make more sense.

All you need to do is create a Google analytics account at http://www.google.com/analytics/. Upon signing up for an account, you will need to do the following to start collecting data from your web pages; simply add the urchin.js file or the ga.js file to all the head tags of every page on your website.

For those of you that are not techies, no problem. Just have your web person do it, it's relatively easy.

Once the urchin is installed, you will start to see the traffic stats. Analytics include where visitors are coming from, which landing page they are going to, and whether or not they are reaching the call to action or goal in Google terms.

However, when this book went to press, Google Analytics was having issues tracking traffic from social media applications, link shorteners, and some mobile devices.

☑ **Tip**: Go to http://www.google.com/analytics/ right now and at least start tracking your main website.

Changing and Tweaking

Obviously, just monitoring your landing pages is pretty useless unless you're prepared to change the bad performers. All landing pages with little to no traffic should be removed and replaced with new ones.

A good practice is to group the landing pages into categories. For example, with Scate Ignite we had four categories:

1. Social media creation
2. E-learning creation
3. Corporate training
4. Selling and marketing

These are then monitored and changed as separate campaigns.

Good landing pages are left while bad ones are replaced with a new message containing related keywords. This process is ongoing and should never stop.

☑ **Tip**: Changing and tweaking landing pages should never stop!

Finally, Real Closing!

During my previous position as Director of Engineering, I never really appreciated the work ethic of some of our sales representatives. As a matter of fact, from my perspective, all they did was golf, facilitate meetings, and push a lot of paper! ☺

Then one day, while presenting a new quote to a customer, it dawned on me that I was actually doing part of the selling. Not only was I compiling many of the presentation packages, but I was also presenting them to the customers.

I was in denial. "I'm not a sales person, I'm the Director of Engineering," I thought to myself. But regardless of my title, I was selling. I just wasn't performing the ASK. It was the account rep that was performing the ASK, but I didn't realize the importance of this until years later. Obviously the reps could not have landed these large contracts without proving to our customer that we had the capabilities to perform and succeed. But I did not appreciate that the account reps' value was in the ASK.

What Is the ASK Anyway?

Why do rich people always seem to get the best deals? It's simple: they are not afraid to ask for one.

"Asking" is the key component to closure and was said best by Terry Bean, a very good friend of mine, in his book, *The Universal Guide to Business Networking.* "Asking is always the predecessor to what you want." (By the way, I highly recommend the book!)

To expand on Terry's quote, it is also the predecessor to what you get and if you are a business owner, you will know exactly what I mean. Sales people need to close and this means they need to ASK and ASK and ASK for the deal.

This is very uncomfortable for many and that is probably why poor people far outnumber rich people. I have heard many people say, "He must have done something illegal to get rich", or "She must have slept with someone to get the part." Not only is this a jealous statement, it is also a true indicator that those people do not know what it takes to become successful.

In many cases all you need to do is ASK. It's nothing new and my daughters, Noelle and Lindsey, always remind me to just look in the *Basic Instruction Before Leaving Earth Book* – The Bible: it has all the answers.

☑ **Tip**: Matthew 7:7 "Ask and it will be given to you; seek and you will find; knock and the door will be opened to you."

☑ **Tip**: Matthew 7:8 "For everyone who asks receives; he who seeks finds; and to him who knocks, the door will be opened."

Steve's 5 Steps to Closing a Deal

OK, let's close the deal. Say your landing pages are all set up and you're tweeting up a storm. So, how do you convert these contacts into sales?

Here is my simple five-step process you can follow every time:

<u>STEP #1</u>: Be Patient. Ask the customer to see the product or service at a meeting or the webinar.

Whatever you're selling, people need to see it first. For example, anyone downloading software from Scate's landing pages is automatically responded to with an email thanking them for downloading and providing them with a coupon. Then a second personalized email is sent asking them to attend a webinar to see the product live in action. The only goal of a sales representative at this step is to get the person to see the product in a webinar or live meeting. They do not try to sell it at this point.

<u>STEP #2</u>: Demonstrate the value proposition at a webinar or live meeting.

During this step you need to provide a solid value proposition by demonstrating the product, explaining its features, comparing its benefits over your competitors, and listening and answering the customers' questions. This shows that you're willing to spend the time and effort to alleviate their pain and support their needs. This is a key part in starting to build a relationship with them.

Many people have asked me, "How can you sell to people over the internet when you never meet them face to face?" My answer is quite simple; you do not need to physically sit beside someone in today's hi-tech world to spend time with him or her. Webinars are very effective and super easy to use. At Scate, we broadcast our webinars from our corporate office many times a week using GoToMeeting located at: www.GoToMeeting.com.

STEP #3: Ask the customer where to send the quote.

This step is simple. Send them a quote and make sure the customer knows exactly what they're getting and for what price. Make sure you have a standard pricing model. Do not quote different rates to different customers for the same thing - it's a small world and people talk.

STEP #4: Negotiate the final price.

If your customer asks for a discount, do not lower your price beyond the standard pricing model, such as coupons, etc. From my experience, when you lower price upon request, you lose all credibility. You do not want the customer to wonder why you didn't give them the best price from the beginning. If a customer wants a price decrease, then the value of your offering must also decrease in the same proportion. If you want your company to stay in business, then always sell by value and not by discount.

It sounds so simple, but negotiating is where I struggle the most. Sales people tend to screw up when emotion gets involved. Just remember that even though you are the one negotiating the sale, the customer on the other end is selling you on getting the best deal. That is their right, so don't get mad, just stick to the process and you will close profitable deals.

STEP #5: ASK the customer where to send the invoice.

Asking where to send an invoice is the easiest method to close after final negotiation of price. Once a customer has a quote, this is the only other step between you and closure. Your goal is to get the invoice into the hands of the customer's accounts payable department.

There is obviously one other step... getting paid! Over the last ten years of business, I could probably write an entire book on that topic alone, so stay tuned.

☑ **Tip**: Follow *Steve's Five Steps for Closing a Deal*.

Section IV: Putting it All Together

Tell Them What You Told Them!

I remember one thing my English teacher at Central Michigan University told me to do when presenting:

1. Tell them what you're going to tell them.
2. Tell them.
3. Tell them what you have told them.

For some reason, humans need to read, see, or hear something at least three times to remember it long-term. Therefore, the following two sections will accomplish the third item from above and recap highlights from this book.

Here We Go!

Here we go, almost everything you learned in one nice neat package...

Once you get a taste of entrepreneurship, the feeling of freedom will stick with you forever. Sure, there will be times that you want to give up. If you fail in business, try again and again. Each try will be a learning experience for you to build upon. I believe God kept telling me to keep trying.

We all have little voices inside our head telling us that we "can't do it" or that "it's impossible." It's all a trick of the mind, so don't listen. Remember your life will fit seamlessly together if you just let it happen and act on opportunities in a timely fashion. The problem is, we can't see the big picture until we sit back later in life and write a book. ☺

Remember life is too short to do a job that you hate or do just for the money. You've got to feel good about what you're doing. Just make sure you finish your education, even if it takes you ten years. Time will pass whether you do or do not finish and it's never too late to learn.

If you decide to start a new business after reading this book, remember that great leaders allow their employees to grow. You should want to help your employees be better than you.

To build a successful company requires humility, patience, strength, trust, faith, commitment, persistence, and loyalty.

Be constantly looking for new methods to help your customers reduce costs. Also be open, available, and nimble enough to jump on opportunities when they are presented. Always look for new markets to arise and ways to remove your customer's pain.

Don't be afraid to get help from others. It is impossible to know everything and that's OK, just don't be too hasty to take investment money.

Remember that the grass is always greener on the other side. So don't get side-tracked on expensive sales trips.

Listen carefully to your customers. They will tell you everything you need to know about your product or service. However, don't be too hasty to jump to conclusions about your offering. Doubt is normal - get over it.

Landing pages are the best method to leverage the web and drive traffic. Just make sure your landing page title, URL, content, and tags all contain the keyword you want to optimize. I definitely advise using landing pages with Yahoo and Google pay-per-click until your landing pages are thoroughly working and indexed. Also, make sure the keywords and subject of the pay-per-click ads contain the keywords for which the landing pages are optimized, as this will lower your cost. Remember to shut off your pay-per-click campaign periodically to stop the cost per click from ratcheting up.

Create marketing presentations with Scate Ignite and broadcast them on Ignitecast.com and YouTube. Post them on Twitter, ScreenTweet, and Facebook. Bookmark in Digg and Delicious. This will provide another level of exposure through cross linking.

Sometimes it's the simple things that make you a success! Twitter is simple and Tweets are like TV or radio spots that require quantity, variety, and frequent broadcasting. If you use landing pages and social media correctly, you can drastically reduce costs - and in some cases completely stop spending money - on other traditional marketing approaches. If you combine social media with Twitter, you have millions of potential clients sitting at your fingertips.

If new technology scares you, try to compare it with something you know. Just remember that opinions do not generate profitability and revenue for business...RESULTS DO!

Don't let naysayers stop you from doing what's right and what works!

Remember that if you can make a profit on a job, then your price is good. Never take on a job for a loss and never feel bad about making a profit. That is what business is all about.

Your price should only change if you find more efficient ways to lower your cost of goods and service or the cost of sales and marketing.

Most people on the internet buy on impulse, so provide visitors to your landing pages with only one or two calls to action. Remove every other button or link and make sure you follow-up immediately, don't procrastinate.

Constant Contact is a great way to consistently follow-up with your contacts and customer base.

If you continue to follow-up with a few offerings for free, you just might create long-lasting customer relationships; just don't de-value your product or service with discounts.

Don't believe that Silicon Valley has the patent on creating new ideas. Just apply what you know to your dreams and remember my five simple steps to closing a deal.

1. Be patient. Don't sell the product; sell the demonstration meeting or webinar.
2. Demonstrate the value proposition of the product or service at a webinar or live meeting.
3. Give them a quote.
4. Negotiate the final price. Remember to sell value. Do not lower the price without proportionally lowering the value of offering.
5. ASK where to send an invoice.

Steve's Quick Tips

☑ **Tip**: Entrepreneurship: Once this feeling of freedom bites you, you're stuck with it forever.

☑ **Tip**: If at first you fail in business, try again. Each attempt is a learning experience for you to build upon.

☑ **Tip**: "Never say 'never'." I would not be where I am today if I had listened to that little voice telling me, "There are never jobs at the unemployment office!"

☑ **Tip**: It took me ten years to complete my degree. So if you're thinking about taking a night class, do it! Time will pass whether you do or not and it's never too late to learn.

☑ **Tip**: If you let things happen and act on opportunities in a timely fashion, you will look back and see how your life fit so seamlessly together.

☑ **Tip**: Good leaders allow their employees to grow.

☑ **Tip**: If you're going to do something do it right or don't do it at all.

☑ **Tip**: Life is too short to stay in a job you hate or do just for the money.

☑ **Tip**: To build a successful company you need to practice humility, patience, strength, trust, faith, commitment, persistence, loyalty, and a whole lot more.

☑ **Tip**: I could not run a successful business without my faith in God.

☑ **Tip**: Business and marketing plans need to be fluid, scalable, and nimble enough to recognize and accommodate new opportunities in the ever changing and uncontrollable world around us.

☑ **Tip**: Look for methods that reduce costs and the corresponding market to sell your product or service.

☑ **Tip**: Be open, available, and nimble enough to jump on opportunities when they are presented. If you don't, some else will!

☑ **Tip**: I don't know what your beliefs are, but if you're praying for opportunities, you better be available and ready to act on them!

☑ **Tip**: Don't be afraid to get help from others. It is impossible to know everything and that's OK.

☑ **Tip**: People are placed in your path for both good and bad purposes, but either way, you will learn from the encounter.

☑ **Tip**: Don't be too hasty to take investment money.

☑ **Tip**: The grass is always greener on the other side.

☑ **Tip**: Listen carefully to your customers. They will tell you everything you need to know about your product or service.

☑ **Tip**: Don't jump to conclusions about your product or offering. Doubt is normal - get over it.

☑ **Tip**: Find a great AdWords training site located at: http://www.perrymarshall.com.

☑ **Tip**: This is the best website to test your html, http://validator.w3.org/, before it goes live.

☑ **Tip**: Make sure your landing page title, URL, content, and tags all contain the keyword you want to optimize.

☑ **Tip**: The more frequently you add or change links in the XML sitemap, the more frequently the robots will index your content.

☑ **Tip**: AddThis and Gigya are excellent social media bookmarking and sharing widgets that you can easily embed on any webpage. You can find them at http://www.addthis.com/ and http://www.gigya.com/

☑ **Tip**: Make sure the keywords and subject of Google and Yahoo pay-per-click ads contain the keyword for which the landing pages have been optimized. This will lower your bid cost!

☑ **Tip**: Create marketing presentations with Scate Ignite and broadcast them on Ignitecast.com and YouTube. Post them on Twitter, ScreenTweet, and Facebook. Bookmark them in Digg and Delicious.

☑ **Tip**: Sometimes it's the simple things that make you a success!

- **Tip**: Tweets are like TV or Radio spots and require quantity, variety, and frequent broadcasting.

- **Tip**: If you do social media right, you can drastically reduce -and in some cases completely stop spending money- on other traditional marketing approaches.

- **Tip**: If you combine social media with Buztweet.com, you have millions of potential clients just sitting at your fingertips!

- **Tip**: If new technology scares you, try to compare it with something you know.

- **Tip**: Opinions do not generate profitability and revenue for business... RESULTS DO!

- **Tip**: Listen to your customer. They will define your value proposition.

- **Tip**: If you can make profit on the job then your price is good.

- **Tip**: Never feel bad about making a profit: that is what business is all about.

☑ **Tip**: Price should only change if you find more efficient ways to lower your cost of goods and service or the cost of sales and marketing.

☑ **Tip**: Provide visitors of your landing pages with only one or two call to action buttons. Remove every other button or link.

☑ **Tip**: Most people on the internet buy on impulse, so follow-up immediately. Don't procrastinate!

☑ **Tip**: Constant Contact is a great way to consistently follow-up with your contacts and customer base.

☑ **Tip**: Giving things away for free can definitely help with sales, just be careful not to de-value your product or service.

☑ **Tip**: You don't have to be in Silicon Valley to be creative, folks. Just apply what you know to your dreams.

☑ **Tip**: Go to http://www.google.com/analytics/ right now and at least start tracking your main website.

☑ **Tip**: Changing and tweaking landing pages should never stop!

☑ **Tip**: Matthew 7:7 "Ask and it will be given to you; seek and you will find; knock and the door will be opened to you."

☑ **Tip**: Matthew 7:8 "For everyone who asks receives; he who seeks finds; and to him who knocks, the door will be opened."

☑ **Tip**: Follow *Steve's Five Steps for Closing a Deal*:

1. Be patient. Don't sell the product; sell the demonstration meeting or webinar.
2. Demonstrate the value proposition of the product or service at a webinar or live meeting.
3. Give them a quote.
4. Negotiate the final price. Remember to sell value. Do not lower the price without proportionally lowering the value of offering.
5. ASK where to send an invoice.

Acknowledgements

Unfortunately I can't mention everyone that has been important to me over the years, so let me apologize in advance for everyone I miss.

Thanks to God, for giving me the talents, patience, and the perseverance to keep on keeping on. I will always be learning till the day I meet you.

Thanks to my father, Alfred Henry Sadler (RIP), for always being there for me. As far back as I can remember, I was in awe of the abilities my father possessed. If it wasn't his amazing creative ability in art, it was his elegance on the dance floor in Gene Kelly proportion. His charisma, charm, and generosity gave him an ability to connect with anyone of any age. He passed in 2003 and I pray that he knows I am appreciative for everything he and my mom sacrificed for me.

Thanks to my mother, Constance Irene Sadler, for always being there and having the courage to move the family to Canada at the age of 55. She also showed great strength throughout her life and survived WWII in England. Dad also knew he would not have been the dancer he was without her.

Thanks to my wife Laurie who has stuck beside me for over twenty (20) years and is still copyediting my poor grammar and spelling. I love you.

Thanks to all my children: Michael, Noelle, and Lindsey. You have been my driving inspiration to succeed since you were born. I am richly blessed to have such talented and caring human beings in my life.

Thanks to my long-time business partners, Jeff Holth and Dan Rea. I thank you guys for your loyalty, dedication, and believing in me. You should both be commended for everything you have done for Scate.

Thanks to Phil Palazzolo (RIP) for always being there and believing in SCATE. You are truly missed.

Thanks to Herman Moore for always having faith in our skills and becoming a good friend to Laurie and I.

Thanks to my old friends and ex-bosses Nick Parker and Mike Veltri. Words cannot describe how appreciative I am for the opportunities you both gave me to grow.

Thanks to my good friend, Tom Bommarito. Whether he knew it or not, he taught me a tremendous amount about sales and developing long-lasting business relationships.

Thanks to all the developers at Scate—you guys are the best and I wish all of you continued success and prosperity.

Thanks to Bonnie and Denny Emmons (Pastor Denny) for being good spiritual mentors to our community.

Thanks to Terry Bean for inspiring me to finish my book and for inviting me to a formal networking event, while telling me, "Jeans are OK to wear."☺

Thanks to Sue Music for copyediting this book and always being there for Laurie and I throughout the years.

Finally, thanks to all my family and friends for just being you, God Bless.

About the Author

Stephen Sadler (Steve) was born in Nottingham England, raised in Blenheim, Ontario, Canada and in 2006 he and his family became US Citizens. Steve and his wife Laurie are the founders of the Michigan-based Company SCATE® Technologies, Inc. (The inventor and developer of Scate Ignite, BuzTweet.com, ScreenTweet.com and IgniteCAST.com social media and e-learning technology).

Prior to SCATE, Steve worked for many years as a Director of Engineering for a large automotive parts manufacturer. Steve is also a summa cum laude science graduate from Central Michigan University and has been honored with awards for leadership and entrepreneurship.

Steve currently holds patents and patents pending in various technologies. He has assisted many other businesses in developing and launching their new technology.

In addition, Steve is an accomplished software and robotics programmer who still enjoys teaching.

On a personal side, Steve was a competitive bodybuilder who now enjoys coaching football (soccer), playing ice hockey, lifting weights, and socializing at his Church (GCC Gingellville Community) with his family and friends.

10 Years of Milestones and Achievements

9/9/10 - Stephen publishes his first book, Exposure to Closure.

6/10/10 - Scate launches BuzTweet.com.

3/19/10 - Steve was nominated for the Ernst and Young *2010 Entrepreneur of Year Award*.

2/25/10 - Scate becomes a Microsoft Development Partner.

2/18/10 - Scate was voted *"Best of Michigan" 2010* by the Corp! Magazine Readers.

10/23/09 - Steve received TiE 2009 Midwest award for Scate's software excellence.

8/31/09 - Scate launches ScreenTweet.com.

4/28/09 - Steve is an honoree of Lawrence Technological University as a *Leader & Innovator Award* recipient.

3/12/09 - Scate launches the expansion of its new web and application development division.

10/30/08 - Scate launches *Scate Ignite 4 Home Edition* for creating social media at a Scate event.

5/21/08 - Scate launches *Scate Ignite 4* at an event.

11/16/07 - Scate launches IgniteCAST.com.

5/1/07 - Scate launches *Scate Ignite Pro-Studio* and *Scate Ignite Expert Edition*.

1/18/07 - Scate signed CAA software development agreement with Dassault Systèmes of France.

7/2/06 - Scate launches the expansion of Scate Technologies, Inc.

6/4/06 - The Sadlers became American citizens.

3/15/06 - The Sadlers were baptized at GCC by Pastor Denny Emmons.

4/26/06 - Scate launches *Scate Ignite 2006*.

2/22/06 - Scate establishes ITV3 video technology.

1/26/06 - Scate launches *Scate MVV – Multi-Vector Vision*.

11/2/05 - Scate releases to the public *Scate Ignite v2.0* at a Scate Open House.

7/29/04 - Steve negotiates contract with Delphi to create advanced design methodology online delivery (e-learning).

4/4/03 - Steve and Laurie founded Scate.

6/30/01 - Steve graduated summa cum laude from Central Michigan University.

Photos

1986 - I still have a shirt from the ladies softball team we sponsored. Forgot to mention I was also the manager / coach.

1989 - Bluewater Classic Amateur Body Building Competition in Sarnia, Ontario. Entered as a lightweight, weighing in at 168 lbs.

1992 - Programming the first robot at my previous employer. Yes folks, they are suspenders. No wonder I'm not smiling! :-)

1992 - Here is a photo of the integration of the weld line. I can't even begin to explain how much I learned during this period.

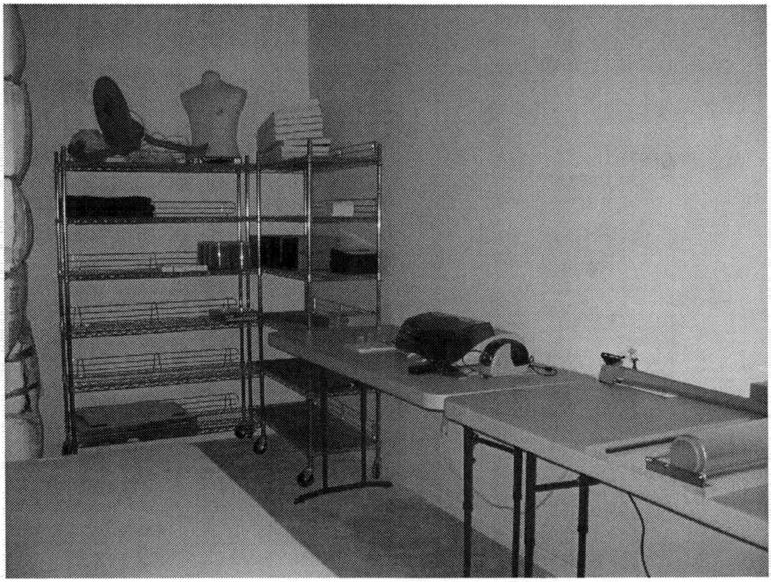

2005 - Settting up the Software Production Room. CD duplicator, software box shrink wrapper. This is now completely obsolete.

2005 - Setting up one of Scate's Hi-Tech development labs in Orion, MI. Computers are waiting for table tops.

2005 - Here is Scate's CFO and my wife Laurie Sadler (Benoit), who I'm very blessed to have as a life and business partner.

2005 - Here are our girls helping to setup Scate's new 7200 sq. ft HQ.

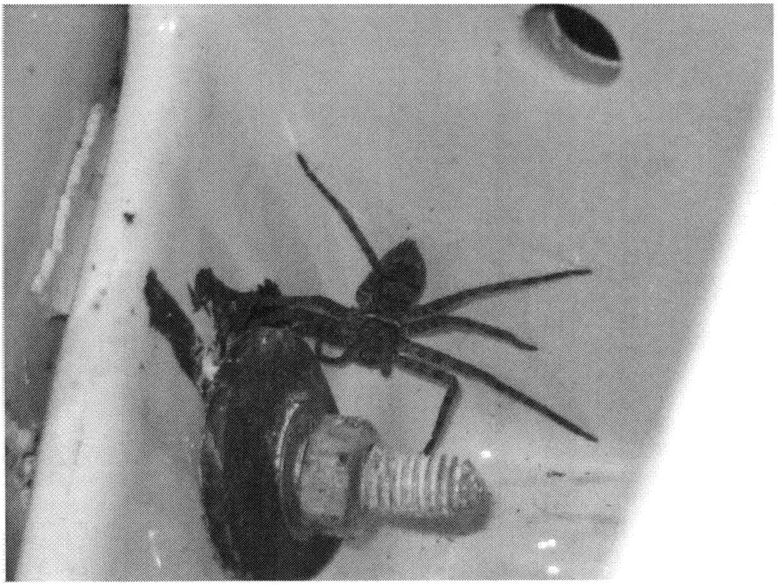

2005 - The infamous spider from Punar India.

2005 - Dan Rea standing beside the old English tea plantation in Punar, India. This is the place I believe almost killed both of us!

2005 - Dan Rea, Oakland County Executive L. Brooks Patterson, Yours Truly & my wife Laurie at the Scate Ignite commerical launch.

2005 - Dan Rea and I at the Scate Ignite commercial launch. My daughter Lindsay says, "Dan Rea and I are posers!" :-)

2007 - Jeff Holth, Chief Technology Officer of Scate Technologies, getting ready to broadcast a Scate Ignite webinar.

2008 - Mad Scientist Jeff Holth demos the new version of Scate Ignite at a users group. Jeff is a smart and loyal friend!

2009 -10 Awards - 1. Lawrence Technological University Leaders and Innovators Award. 2. Corp Award. 3. TieCon Technology Award.

2010 - Dan and I cooking at the Dave Thomas Spaulding for Children Celebrity Cookoff at Benihana in Troy, MI. Great event, great cause.

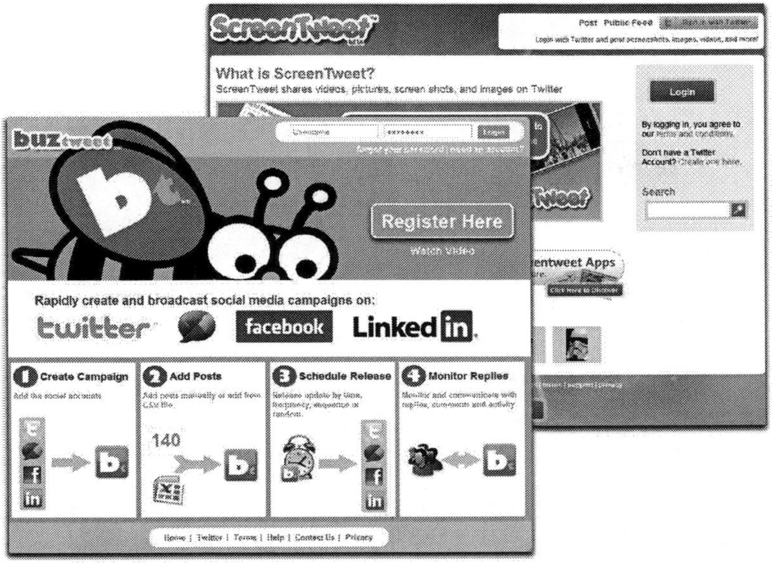

http://screentweet.com for free social media posting (photos, video)
http://buztweet.com for creating social media marketing campaigns

http://scateignite.com for creating social media presentation & video
http://ignitecast.com for broadcasting Ignitables made in Scate Ignite

2010 - My wife took this shot.
In the Big D...Christ covers ALL folks...Awesome!